www.trishhart.com

upside down
turn around
SERIES

Trish Hart

upside down
turn around

BIG
small

Trish Hart

www.trishhart.com

upside down turn around SERIES

First published in 2004 by Funtastic Publishing

National Library of Australia Cataloguing-in-Publication entry
Creator: Hart, Trish, 1956- author, illustrator.
Title: Big and small : a concept picture book for children / Trish Hart.

ISBN: 9780648011217 (paperback)
Series: Hart, Trish, 1956- Upside-down turn-around ; Vol 1.

Target Audience: For primary school age.
Subjects: Animals--Juvenile literature.
Animals--Pictorial works.
Picture books for children.

Published by Hartpix Press
2 Island View Road
The Gurdies, Victoria, 3984 Australia

Some animals are bigger than me!

BIG

Read this book with a friend facing you
or
read all the **small** pages,
turn,
then read all the **big**
or
read and turn every page!

We found some animals that are fat,
some that are tall,
and some that are just bigger than us.

Read this book with a friend facing you
or
*read all the **big** pages,*
turn,
*then read all the **small** pages*
or
read and turn every page!

small

Some animals are smaller than me!

Pigs are **big** animals
that like to roll in mud.
This keeps them cool and
stops them from getting sunburnt.
Pigs grunt and squeal and
sometimes they are a bit smelly.

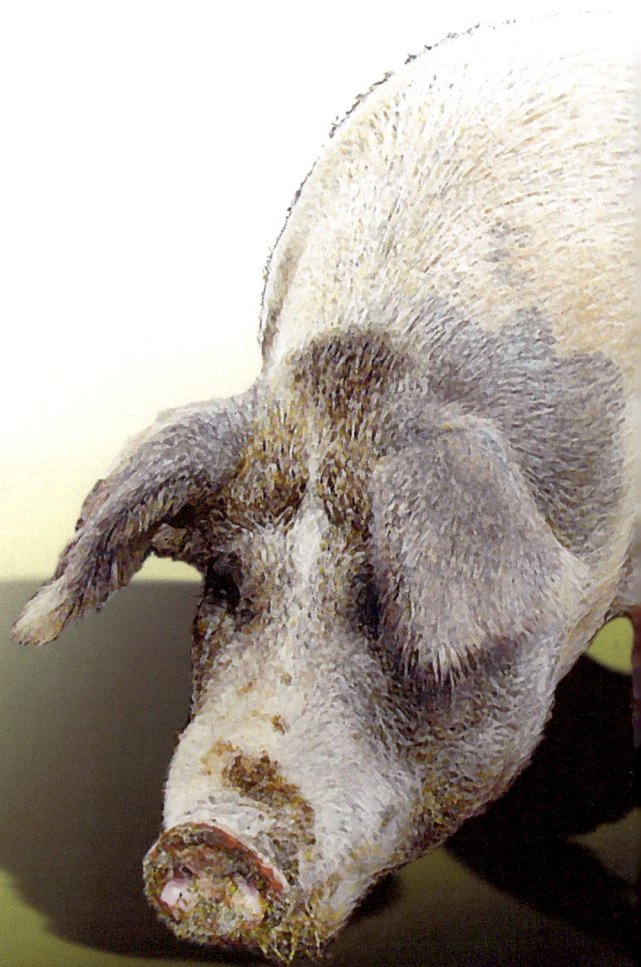

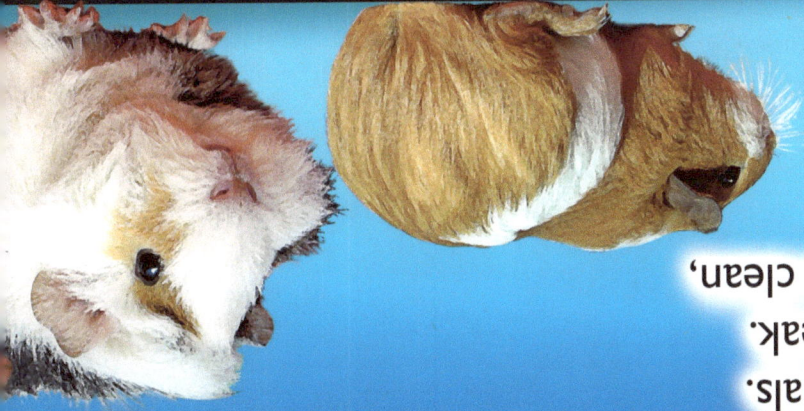

Guinea pigs are **small** animals.
They grunt, squeal and squeak.
If you don't keep their cage clean,
they are a bit smelly.

Crocodiles are **big** animals.
They are very quick and strong.
They wait in the river for a feed.
Look out!
You could disappear
in one gulp.

Geckos are **small** animals.
They can walk on the ceiling and
swallow small bugs in one gulp.

Whales are **big** marine animals.
They are the biggest animals in the world.
Some like swimming amongst the icebergs
in the freezing cold oceans
of the Antarctic.

Clownfish are **small** marine animals. They like to swim in the warm waters of tropical reefs amongst the tenticles of anemones.

Rhinoceroses are **big** animals. They are very strong and use their big horns to fight. Their horns are made of keratin - that's the same stuff as your hair and nails.

Rhinoceros beetles are **small** animals.
They use their big horns for battle and are
the strongest creatures in the world.
They can lift more than
100 times their own weight.

Zebras are **big** animals.
When they stand together
it is hard to see where one zebra ends
and the next one begins.

Zebra fish are **small** marine animals.
When they swim together
it is hard to see where one fish ends
and the next one begins.

Some wildcats are **big** animals.
These pumas are very strong
and pounce on their prey.
You wouldn't want to
have a puma as a pet.
They have very big teeth
and sharp claws.

Our pet cats are **small** animals.
They can have all sorts
of patterns and colours.
Do you have a cat for a pet?

Horses are **big** animals.
They are so big and strong
they can carry people
on their backs.
Have you
ever ridden
a horse?

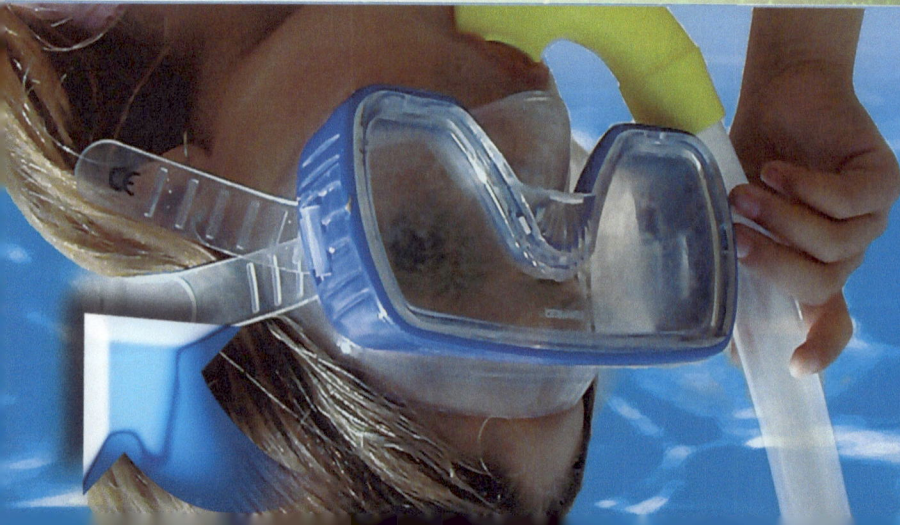

Seahorses are **small** marine animals.
They live in shallow sea water.
To ride a seahorse you would have
to be smaller than a mouse.

Peafowls are **big** birds.
The males are called peacocks
and have beautiful coloured
feathers so they can show off
to the females.

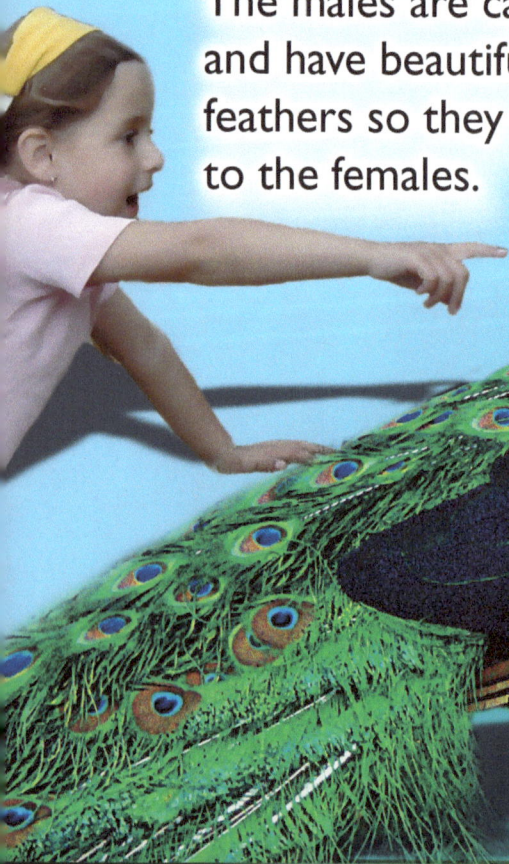

Wrens are **small** birds.
The males have beautiful
coloured feathers so they
can show off to the females.

Elephants are **big** animals.
They are too big to keep
as pets in your bedroom.
Did you know that
elephants can't jump?

Mice are **small** animals
that make good pets.
They are smaller than an
elephant's little toe.

Kangaroos are **big** animals
that are good at the long jump.
If you lift a kangaroo's tail off
the ground it can't jump.
They need their tails for balance.

Pet rabbits are **small** animals and come in all sizes and colours. Rabbits can hop very fast.

Gorillas are **big** animals.
They live in family groups and
like to walk on all fours.
They can chuckle, smile,
and burp.

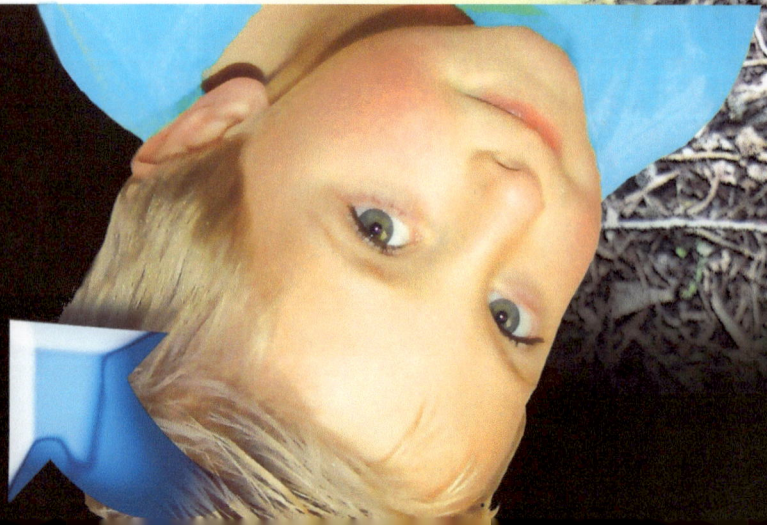

Kids are **small** mammals.
They live in families
and like to play.
Kids often laugh, smile,
and burp.

Some animals are bigger than me!

BIG

Read this book with a friend facing you
or
read all the **small** pages,
turn,
then read all the **big**
or
read and turn every page!

We found some animals that are fat,
some that are tall,
and some that are just bigger than us.

Read this book with a friend facing you
or
read all the **big** *pages,*
turn,
then read all the **small** *pages*
or
read and turn every page!

small

Some animals are smaller than me!

Thanks

to all the animals who posed for my book-
Faye's bunnies and guinea pigs,
Sally, Liz's horse,
the pigs that belong to Aarons dad, John,
Cheeko, Beth's burmese cat,
Maddison and Connor (children), Zipper and Babbs (their cat and rabbit),
Phew...
Oh, and thanks to Bob from the Gould League,
Peter and the gorilla from Melbourne Zoo, Victoria,
Roos and crocodiles from Ballarat Wildlife Park, Victoria,
ReefHQ, Townsville,
Whipsnade Wild Animal Park, U.K,
and Linda, Chris, and Liz.

Hope you enjoyed going upside down and turning around.
Look out for more in the **upside down, turn around** series.

It is something that belongs to the sea.
*Do you know what **marine** is?*

It is a warm blooded animal that produces milk to feed her baby.
*Do you know what a **mammal** is?*

www.trishhart.com

upside down
turn around

BIG small

Trish Hart

the other books in the upside down turn around series are

Antarctic and Arctic

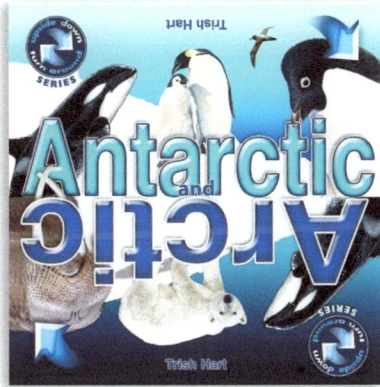

Read the book one way to see the Antarctic animals, or turn it upside down to see ones that live in the Arctic!

Wet and Dry

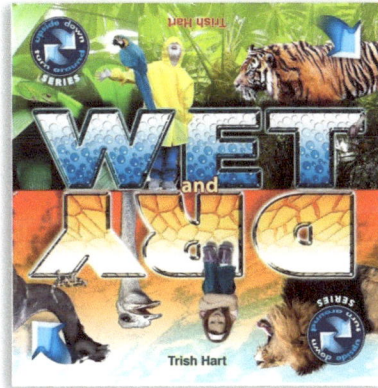

Read the book one way to see the animals who live in wet places, or turn it upside down to see those who love the desert!

Day and Night

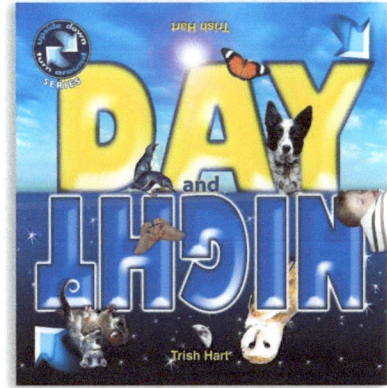

Read the book one way to see the daytime animals, or turn it upside down to see the ones that come out at night!

Trish Hart

upside down
turn around

BIG
small

Trish Hart

Thanks

to all the animals who posed for my book-
Faye's bunnies and guinea pigs,
Sally, Liz's horse,
the pigs that belong to Aarons dad, John,
Cheeko, Beth's burmese cat,
Maddison and Connor (children), Zipper and Babbs (their cat and rabbit),
Phew...
Oh, and thanks to Bob from the Gould League,
Peter and the gorilla from Melbourne Zoo, Victoria,
Roos and crocodiles from Ballarat Wildlife Park, Victoria,
ReefHQ, Townsville,
Whipsnade Wild Animal Park, U.K,
and Linda, Chris, and Liz.

Hope you enjoyed going upside down and turning around.
Look out for more in the **upside down, turn around** series.

Do you know what **marine** is?
It is something that belongs to the sea.

Do you know what a **mammal** is?
It is a warm blooded animal that produces milk to feed her baby.

www.trishhart.com

Trish Hart

My name is Connor.

BIG

small

My name is Maddison.

Trish Hart

www.ingramcontent.com/pod-product-compliance
Lightning Source LLC
Chambersburg PA
CBHW042049210326
41520CB00043B/187